Thiago Leite de Alencar

Characterization and impacts of

GW01403151

Thiago Leite de Alencar
Márcio G.R. Lobato
Sebastião C. Sousa

Characterization and impacts of different uses of an Argissolo

A study in the semi-arid region of Northeast Brazil

ScienciaScripts

This book is a translation from the original published under ISBN 978-3-330-75498-0.

Publisher:
Sciencia Scripts
is a trademark of
Dodo Books Indian Ocean Ltd. and OmniScriptum S.R.L publishing group

120 High Road, East Finchley, London, N2 9ED, United Kingdom
Str. Armeneasca 28/1, office 1, Chisinau MD-2012, Republic of Moldova, Europe

ISBN: 978-620-8-26925-8

CONTENTS

1 INTRODUCTION

Soil management is currently one of the factors that most positively or negatively influences agriculture, because depending on how it is used, the degree of soil degradation is variable, and this is a limiting factor for the full development of agricultural activities.

Improper use of the soil causes various environmental damages, such as erosion, compaction, increased salinity, reduced biodiversity, silting of water bodies and reduced soil fertility. Therefore, the importance of proper management in agricultural activities is clear, so that rural development can take place in a sustainable manner.

In the Cariri region, soils are used in a variety of ways, from conservation to inadequate use, mainly through the use of large numbers of animals in the area and conventional planting. The area used for this work is located in the municipality of Assaré - CE, which belongs to the Cariri region of Ceará.

Soil has a variety of fundamental functions for the balance of an ecosystem, with physical functions that act in the development and support of the root system of vegetation, the supply of nutrients and o_2 for plants and the availability and supply of water, therefore, soil must be used and managed in an appropriate way that maintains the ability to perform its functions.

Different soil management systems can have a direct influence on the degree of compaction, the rate of surface runoff, resistance to penetration, soil density, porosity and the amount of water available for plants.

Therefore, the importance of a study that characterizes the influence of different managements on the physical and chemical attributes of the soil is evident, in order to determine which model is the least degrading and most appropriate in terms of soil quality and sustainability.

The aim of this work was to evaluate the influence of different land uses on the physical attributes of an Argissolo located in the semi-arid region of northeastern Brazil, subject to critical water deficiencies and inadequate management techniques by the landowner.

2

2. LITERATURE REVIEW

2.1. Physical attributes of the soil

2.1.1. Compaction

According to Collares *et al.* (2006), compaction alters the structure of the soil, leading to an increase in penetration resistance, soil density and a reduction in macroporosity, thus affecting plant growth and root development.

Compaction is generally found in soils with inadequate management, generating major economic and productive losses, especially in regions with irregular rainfall distribution (GUIMARÂES *et al.*, 2002).

According to Freitas (1994), soil compaction is the factor that most limits the high productivity of crops all over the world, as it directly affects the growth and development of roots, reduces the infiltration capacity of water in the soil and reduces the translocation of nutrients, resulting in a small layer for the roots to exploit.

The differentiation of compacted layers found in the field and the characterization of soil behaviour in relation to its physical properties, such as porosity, infiltration rate and density, is of fundamental importance for the development of production (LAIA *et al.*, 2006).

The process of compaction occurs in a three-dimensional system, due to the influence of mechanical stresses commonly caused by the traffic of machinery and the action of agricultural implements (FLOWERS ; LAL, 1998) and or the trampling of animals (BHARATI *et al.*, 2002).

A qualitative and quantitative diagnosis (degree of soil compaction) is of great importance, not only to help verify the quality of the management used, but also to help establish compaction limits that do not affect the root development of plants in the different management systems (TAVARES FILHO *et al.*, 2001).

Because compaction is a serious problem in agricultural soils, it is being studied intensively (ALMEIDA *et al.*, 2008).

The physical attributes of the soil used to assess the degree of compaction include: soil density, resistance to penetration, macroporosity, water infiltration and, more recently, pre-consolidation pressure (PACHECO ; CANTALICE, 2011).

2.1.2. Penetration resistance

Soil resistance to penetration is one of the most important physical properties for managing and studying the physical quality of soil, as it is directly linked to various soil attributes that indicate the level of compaction (RIBON ; FILHO, 2008). It is a mechanical impediment that the soil offers to root development (SILVA et al., 2008).

It is necessary to know the levels of compaction that reduce the growth and development of the plant root system, with a view to reducing the adverse effects of compaction and using the soil efficiently and sustainably (BERGAMIN et al., 2010).

Periodic monitoring of the level of soil compaction through the penetration resistance factor is a practical and feasible way of assessing the effects of different managements on soil structure, growth and root development of different crops (FILHO ; RIBON, 2008).

Soil resistance to penetration is strongly influenced by a number of factors, such as the soil's texture, structure, porosity, density and mineral composition (BEUTLER et al., 2007). Soils with high values of resistance to penetration make it difficult for water to infiltrate and percolate into the soil, thus increasing surface runoff and erosion.

Soil penetration resistance (PR) is generally determined using penetrometers, the main advantages of which are the ease and speed with which the results are obtained (BLAINSKI et al., 2008).

The penetrometer models most commonly used to determine the degree of compaction are: static penetrometers (which record the PR per unit area), dynamic or impact penetrometers (which record the PR per unit depth) and electronic penetrometers (which record the PR based on the soil's electrical resistance) (ARAUJO, 2010).

The impact penetrometer has been widely used in the field to characterize compaction caused by soil use and management (TORMENA ; ROLOFF, 1996; CASAGRANDE,

2001) due to its low cost, the fact that it does not require frequent calibration and the fact that the results are independent of the operator.

2.1.3. Soil density

Soil density is a physical property that depends on pedogenetic factors and processes in uncultivated environments and is one of the most widely used properties for assessing soil compaction (REINERT *et al.*, 2008; MONTANARI *et al.*, 2010).

To identify compacted layers caused by different cultivation systems, soil density (Ds) is used as an indicator of the degree of soil compaction (CAMARGO ; ALLEONI, 1997).

It is a property that generally varies between and within different types of soil and depends on a number of factors, such as the degree of compaction, organic matter content, the presence or absence of vegetation cover, the cultivation system used and depth (THIMÓTEO *et al.*, 2001).

Soil density is one of the parameters that control air-water relations and indicates both the state and prospect of root penetration, as well as guiding soil management (JORGE *et al.*, 1991).

Different methods are used to determine soil density, with the volumetric ring method being considered the most widely used standard, consisting of sampling soil with an undeformed structure in a ring (metal cylinder) of known volume (PIRES *et al.*, 2011).

The soil density obtained by the destructive method (volumetric ring), as well as being an indicator of soil quality, is used to determine the amount of water and nutrients in the soil profile based on volume (MENDES *et al.*, 2006).

According to Llanillo *et al.* (2006), soil density values above 1.65 Mg m^3 in medium-textured soils are restrictive to the development and growth of the root system and the consequent yield of cultivated species.

2.1.4. Macroporosity

According to Corsini *et al.* (1999), macroporosity combined with soil density can

5

assess total porosity and pore size distribution, which are physical characteristics of the soil indirectly related to structure.

Macroporosity, in particular, is one of the physical attributes used as an excellent indicator of soil degradation, due to its relationship with compaction, since porosity regulates various ecological properties of the soil (STOLF *et al.*, 2011; REICHERT *et al.*, 2009).

Macroporosity is the volume of pore space in the soil where gas exchange takes place (AMARO FILHO *et al.*, 2008). The changes caused to porosity due to inappropriate use of the soil not only alter gas exchange rates, but also the availability of water for plants (ARGENTON *et al., 2005)*.

Lipiec *et al.* (1991) observed a correlation between the degree of compaction and the penetration resistance and aeration porosity of the soil.

The degree of compaction corresponding to the critical macroporosity values depends on the type of soil. In Latosols, the critical aeration limits are reached with a lower degree of compaction than in Argissols (SUZUKI *et al.*, 2007).

According to Pauletto *et al.* (2005), low values of macroporosity and high values of the micro/macro pore ratio imply poor aeration in the soil, directly affecting the full development of rainfed crops.

Soil aeration is a process that involves the exchange of gases, especially carbon dioxide and oxygen, between the pore spaces of the soil and the atmosphere (AMARO FILHO *et al.*, 2008).

Land use directly influences soil structure. Normally, forest and native field soils have greater macroporosity than other uses (ALBUQUERQUE *et al.*, 2001).

2.1.5. Water infiltration into the soil

Plant production has a direct relationship with the dynamics of water in the soil, so its knowledge is of fundamental interest for any decision on the agricultural use of soils. The method used to monitor water infiltration in the soil is the concentric ring infiltrometer (CALHEIROS *et al.*, 2009).

6

Soil infiltration is the process by which water enters the soil through its surface. Generally, the entry of water into the soil, depending on the wetting of the profile, decreases with time and assumes a constant value called the basic infiltration rate (POTT & DE MARIA, 2003).

It is considered one of the most important processes that make up the hydrological cycle, as it plays a decisive role in the availability of water for crops, groundwater recharge, surface runoff and soil and water management (CECiLIO et al., 2003).

Water infiltration is one of the factors that best reflects the soil's internal physical conditions, since good structural quality leads to a distribution of pore size, which favors the growth and development of the plant root system and the capacity for water infiltration into the soil (ALVES; CABEDA, 1999).

Several factors affect the infiltration capacity of water in the soil, the main ones being: time, initial humidity, porosity and texture, soil density, hydraulic conductivity, among others (VIEIRA ; KLEIN, 2007).

According to Cichota et al. (2003), the variable water infiltration rate in the soil is defined as the volume of water that penetrates the surface unit per unit of time. This is of agronomic importance because of its role in the formation of runoff, due to surface runoff being an erosive agent, and in determining viable irrigation rates, the concentric ring infiltrometer method was used to monitor the water infiltration rate in the soil.

The different types of management and land use alter the physical characteristics of the soil, causing a decrease in the rate of water infiltration into the soil, with a consequent increase in surface runoff rates (PANACHUKI et al., 2011).

With soil compaction, the role played by water in the soil is altered, resulting in reduced infiltration and increased surface runoff, which leads to an increase in the degree of soil erosion (GIRARDELLO et al, 2011).

2.2. Land use and occupation

2.2.1.. Argisols

According to the Brazilian Soil Classification System (EMBRAPA, 2006) and the Exploratory Survey - Reconhecimento de Solos do Estado do Cearà - Vol. 1 (MA. SUDENE, 1973), this class comprises soils with a textural B horizon made up of mineral material with low clay activity, due to the soil material being made up of sesquioxides, clays of the 1:1 group (kaolinites), quartz and other materials resistant to weathering, or high combined with low base saturation or alitic character. Most of the soils in this class show a clear increase in clay content from the surface horizon to the B horizon, vary in depth, are reddish or yellowish in color, are not hydromorphic, are strongly to moderately acidic, with high or low base saturation, occur predominantly on flat and undulating terrain, with hypo- and hyperxerophilous caatinga vegetation and forest/caatinga transition. These soils can be used to grow perennial crops adapted to the region's climate.

2.2.2. Soil quality

Soil quality in different types of ecosystems under the influence of different managements is defined as the ability to maintain its productive potential, biodiversity, together with the increase in the quality of groundwater, surface water and soil air (Karlen et al., 1997).

Recently, there has been a growing concern with controlling soil quality, which has been constantly monitored through the quantification of changes in its attributes due to inadequate use and management systems (NEVES et al., 2007).

The practices used in agricultural activities cause changes in soil attributes, which can lead to a significant reduction in quality, directly affecting the sustainability of the environment and the economic balance of the activity developed (NIERO et al., 2010).

The quality of soil subjected to different crops and management can be assessed through its chemical, physical and biological properties (Shukla et al., 2006).

Organic matter and, consequently, microbial biomass play a direct role in maintaining

8

the productivity of agricultural and natural ecosystems and is one of the factors responsible for soil quality (Gama-Rodrigues & Gama-Rodrigues, 2008).

2.2.3. Land use and management

The different uses and managements of the soil, the intensity and time of use cause various alterations in the soil properties, which may present different levels of degradation between similar soils (ROTH ; PAVAN, 1991; CASTRO FILHO et al., 1998).

Soil and crop management directly influence the state of compaction and the physical characteristics of the soil that are ideal for the full development of plants (COLLARES et al., 2008).

The degrading effects of inadequate management of the soil's natural resource can be seen in analyses of the rate of water infiltration, reduction in macroporosity, surface runoff and dragging of soil particles (SOARES et al., 2005).

The changes resulting from the different types of land use in the semi-arid region, which has peculiar soil and climate characteristics, must be studied in order to propose sustainable models that maximize production and avoid the degradation of soils and natural resources (CÔRREA et al., 2009).

Soils under cultivation cannot maintain their original physical characteristics, but they should be managed in such a way as to alter these characteristics as little as possible, especially those that affect water infiltration and retention, such as porosity, density and aggregation, in order to maintain the sustainability of the system (KAMIMURA et al., 2009).

Understanding and quantifying the impact of soil use and management on its attributes and physical quality is of fundamental importance for the development of sustainable agricultural systems (MATIAS et al., 2009).

2.2.4. Pasture

Inadequate management of natural pastures over many years leads to soil degradation

9

and accelerated soil erosion (CARVALHO *et al.*, 2009).

Excessive animal loads on pastures can affect some of the soil's physical properties, increase its susceptibility to water erosion and reduce its productive capacity (BERTOL *et al.*, 1998).

Excessive use of pasture, through excessive grazing intensities, has caused loss of vegetation cover, invasion of undesirable species, soil erosion and negative environmental impact (GONÇALVES, 2007).

According to the results obtained by Trein *et al.* (1991), after the use of a large number of animals in the area over a short period of time, there was an increase in soil resistance to penetration, a significant reduction in water infiltration into the soil and a decrease in macroporosity in a Red Argissolo cultivated with winter pasture.

The physical conditions of the soil can be altered dramatically by animal trampling all over the surface and sometimes repeatedly in the same place, thus hindering the development and growth of the root system (LEÂO *et al.*, 2004).

The degree of compaction due to cattle trampling is influenced by various factors, such as soil texture (CORREA & REICHARDT, 1995), the grazing system used (LEÂO *et al.*, 2004), the amount of plant residue on the soil (BRAIDA *et al.*, 2006) and high soil humidity (BETTERIDGE *et al.*, 1999).

2.2.5. Corn

Brazil is one of the countries in which maize has a high production potential, reaching 10 t ha^{-1} of grain under experimental conditions and by farmers who adopt appropriate technologies (CARVALHO *et al.*, 2004).

Because it plays a significant role in national grain production, corn stands out as a direct influence on the Brazilian economy (SANTOS *et al.*, 2006). One of the most important agricultural activities in Brazil is corn farming, especially in the Northeast region (DoVale *et al.*, 2011).

In soils used for maize farming, the pressure exerted by machinery traffic on the soil surface during tillage operations normally increases soil density and decreases total

10

porosity, especially macroporosity, leading to an increase in the degree of compaction (TSEGAYE ; HILL, 1998).

According to results obtained by Bergamin *et al.* (2010), additional compaction negatively influences the root system of maize, reducing both its growth and development and causing yield losses.

2.2.6. Native vegetation

The Caatinga is the main biome in Brazil's Northeast region, occupying around 845 km^2 or 54.53% of the region's area (IBGE, 2005), and is of significant socio-economic, ecological and environmental importance.

The northeastern semi-arid region is covered by a large area of the caatinga biome and part of this, corresponding to hundreds of thousands of hectares, is cleared every year to use the area for shifting cultivation (MELO *et al.*, 2008).

The use of soil with native vegetation leads to a large reduction in the variation of soil attributes when compared to agricultural uses and, for this reason, native vegetation is a benchmark for evaluating soils incorporated into agricultural systems (CORRÊA *et al.*, 2009).

The soil with native forest, as it is a preserved area, has a lesser impact on the soil's physical characteristics, with emphasis on greater macroporosity and less resistance to penetration, thus generating an environment suitable for normal plant growth and development (ANDREOLA *et al.*, 2000; ALBUQUERQUE *et al.*, 2001).

Soil degradation begins with the removal of natural vegetation and is accentuated by subsequent cultivation, removing organic matter and nutrients that are not replenished in the same proportion over time and altering their physical properties (SOUZA ; MELO, 2003; BERTOL *et al* 2004).

11

3. METHODOLOGY

3.1. Study area

The Redonda farm is located in the municipality of Assaré, on the western side of the Araripe plateau, close to the border with the municipalities of Farias Brito and Altaneira, in the Cariri region, south of the state of Ceará (Figure 1), 22 km from the municipality's headquarters, with an area of 328 ha, located between the UTM coordinates of longitude 416000 to 421000 and latitude 9235500 to 9239000, at an elevation of 616 m. This study was carried out between November 2011 and June 2012. Figure 1 shows the location of the area where the study was carried out.

Figure 1 - Location diagram of the study area.

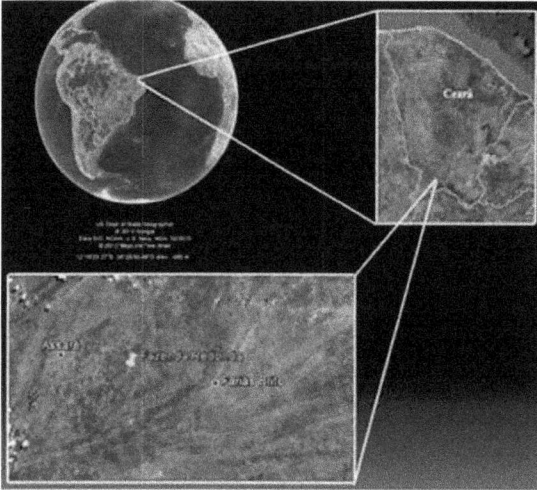

Source: Thiago Leite de Alencar

3.1.1. Characterization of the area

3.1.1.1 Geology

Geologically, according to figure 2, the municipality of Assaré is made up of rocks from the Cretaceous period, belonging to the Araripe series, which comprises the entire Araripe plateau. The area is located in the Exu Formation, which is made up of layers of siltstones and claystones that overlie the clayey, permeable sandstones, with varying granulometry and colors that vary according to the alteration and degree of weathering of the areas, being light (white and gray) in the unaltered parts and red in those where weathering has taken place (MA. SUDENE, 1973).

Figure 2 - Geological map of the state of Ceará

Source: (MA. SUDENE, 1973).

3.1.1.2 Geomorphology

The geomorphological unit of the area is classified as the Araripe plateau, a sedimentary platform originating from the Cretaceous period, with a table-like shape, in which there is a predominance of flat relief, but there are also gently undulating parts and undulating stretches. The slopes of the platform give it the visual appearance of a wall (MA. SUDENE, 1973).

3.1.1.3 Soils

The soils found in Assaré according to the Levantamento Exploratório - Reconhecimento de Solos do Estado do Ceará - Vol. 1 (MA. SUDENE, 1973), they belong to the TRe association, which encompasses different types of soils, among

13

which are the Similar Eutrophic Structured Terra Roxa, corresponding in the classification of (EMBRAPA, 2006) to Nitossolos, Equivalent Eutrophic Yellow Red Podzolic, which belongs to the Argissolos class, Eutrophic Litholic Soils, which are the Litholic Neosols and Eutrophic Dark Red Latosols. The first three types of soil are generally found on slopes with steep terrain. In areas where the erosion process has been more intense, the Neossolos Litólicos soils are concentrated, and both in areas with flatter relief and on higher coasts, the Latossolos.

The distribution of the different soil classes in this association is organized as follows: 40% Nitossolos, 30% Argissolos, 15% Neossolos Litólicos and 15% Latossolos.

3.1.1.4 Vegetation

The caatinga, the area's characteristic biome, is an arboreal-shrub formation, where leaf deciduousness occurs, and has woody formations of varying size. The vegetation is directly influenced by the climate, which is characterized by limited precipitation, uneven rainfall distribution and a prolonged dry period.

Within the biome, the vegetation is classified as hypoxerophilous caatinga due to its less dry climate with dry periods ranging from 5 to 7 months and because it has larger species and is usually denser.

The most common species found are: *Caesalpinia pyramidalis* (catingueira), *Mimosa caesalpinifolia* Benth. (sabià), *Pithecolobium diversifolium* Benth. (jurema branca), *Cassia excelsa* Schrad. (canafistula), *Mimosa nigra-* Hub. (black jurema), *Caesalpinia ferrea* Mart, (pau e ferro), *Pityrocarpa sp.* (catanduva), *Croton sp.* (marmeleiro), Euphorbiaceae; *Combretum leprosum* Mart, (mufumbo), Combretaceae; *Ziziphus joazeiro* Mart, (juazeiro), Rhamnaceae; *Astronium sp.* (aroeira). To characterize the vegetation, we used the analytical information in the Exploratory Survey - Soil Reconnaissance of the State of Cearà - Vol. 1 (MA. SUDENE, 1973).

3.1.1.5 Weather

The region's climate according to the Koppen classification is BSw'h', characterized as semi-arid, with high temperatures, which vary annually between 24 and 26°C and, in

14

the hottest months, between 26 and 28°C (MA. SUDENE, 1973). The average annual rainfall in the region is 660 mm, distributed over a rainy season that runs from December to May. The rainfall figures recorded for the period between 1990 and 2010 show the variations that occur from year to year.

Figures 3 and 4 show the average monthly rainfall between 1990 and 2010 and the annual rainfall at the Assaré data collection point, respectively.

Figure 3 - Average monthly rainfall (mm) (Assaré)

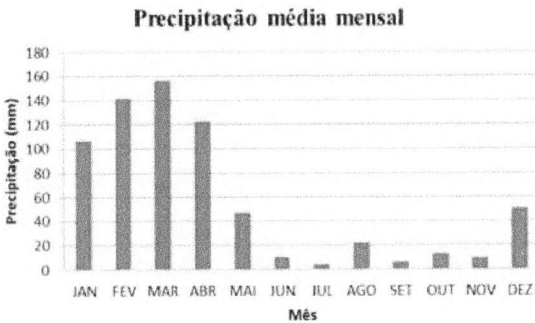

Source: Funceme

Figure 4 - Annual rainfall (mm) (Assaré)

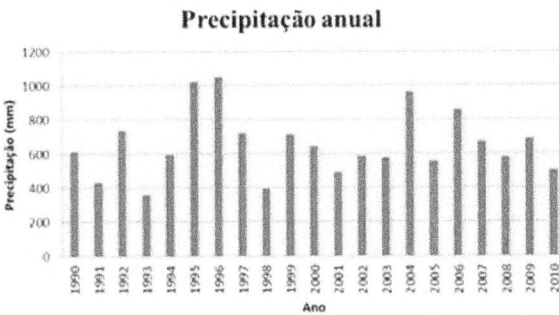

Source: Funceme

3.2. Areas of study

3.2.1. Corn

The area, which has been used for the last 20 years or so to grow corn in a conventional

15

tillage system based on the use of agricultural machinery, was previously used mainly to grow cotton. This area is also grazed by the farm's beef and dairy cattle, an average of 150 animals, which feed on the crop remains from each plantation. Figure 6 shows the corn area used in the experiment.

3.2.2. Pasture

The pasture area (figure 7), which comprises 14 ha, is used for grazing by 150 cattle on average every month of the year. It is important to note that the area has never been used for any other type of agricultural activity, so the soil has never been subjected to the action of machinery.

3.2.3. Native forest

The area with native forest (figure 8) comprises 3 ha and corresponds to the farm's preservation area (legal reserve), consisting of species native to the hypoxerophilous caatinga.

With the help of a navigation GPS, it was possible to georeference (UTM, SAD 69, 24M) the areas of the different land uses, using the ArcView GIS 3.2 program (Figure 5).

Figure 5 - Georeferenced map of the collection areas

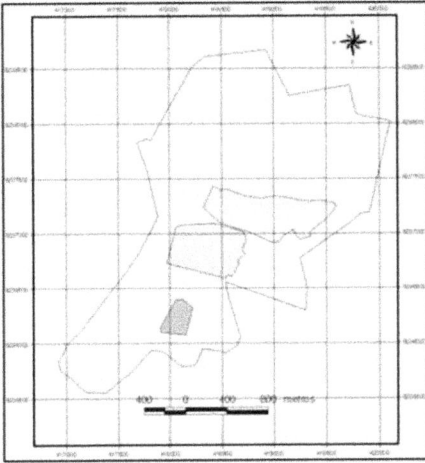

Milho
Pastagem
Mata nativa

Figura 6   - Land use with corn.

Figura 7   - Land use with pasture.

17

Figura 8   - Land use with native forest

Source: Thiago Leite de Alencar

3.3. Soil classification

In order to classify the soil in the study area, whose source material is sandstone, it was necessary to morphologically characterize the soil by collecting samples, which were then analyzed in the laboratory of the Soil Sciences Department at the Agricultural Sciences Center of the Federal University of Ceará - PICI Campus.

The process of morphological characterization of the profile analyzed (figure 9) with subsequent sample collection was subdivided into a number of stages: opening the trench (representative area of the soil under study); differentiating the horizons by observing some important aspects such as soil color, texture, granulometry and structure contrast; separating the properly differentiated horizons and measuring their thickness using a knife and tape measure, respectively; photographic recording of the profile; collecting the samples, which were packed in plastic bags, properly labeled and taken to the laboratory for analysis.

Figure 9 - Profile of the soil analyzed.

Source: Thiago Leite de Alencar

The morphological analysis of the profile in the field followed the procedures set out in the Manual for the description and collection of soil in the field (SANTOS et al, 2005).

3.4. Laboratory analysis

After collecting the soil samples, they were sent to the Soil Chemistry and Physics Laboratories of the Soil Sciences Department at the PICI Campus of the Federal University of Ceará in Fortaleza, where the chemical and physical analyses were carried out respectively.

The analysis of the samples for soil classification was based on the methodologies in the Manual of Soil Methods and Analysis (EMBRAPA, 1997). The Brazilian Soil Classification System (EMBRAPA, 2006) was used to classify the soil up to the 5th categorical level according to the chemical and physical results obtained.

The samples for analysis in the laboratory were prepared according to the flowchart shown in Figure 10.

Figure 10 - Flowchart for preparing soil samples for analysis.

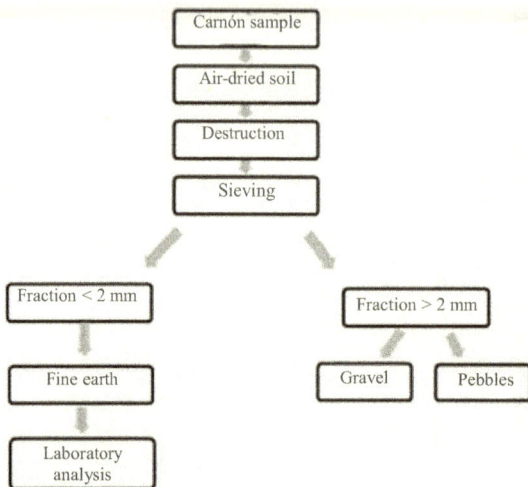

Source: (SILVA, 2010).

3.4.1. Physical analysis

Granulometry

Particle size analysis was carried out using the pipette method according to Embrapa's Soil Analysis Methods manual (EMBRAPA, 1997). Sodium hydroxide (NaOH) was used as a chemical dispersant for the analysis. In this procedure, a volume of the suspension is pipetted to determine the clay, which is dried in an oven and weighed; the sand is separated, dried in an oven and weighed; the silt corresponds to the complement of the 100% percentages.

Particle density

This involves determining the volume of ethyl alcohol needed to complete the 50 ml capacity of a volumetric flask containing oven-dried soil.

Initially, samples containing 20g of soil were stored in tins of known weight and placed in an oven at 105°C for a period of 6 to 12 hours, after which they were dried, weighed and transferred to a volumetric flask. Ethyl alcohol is added until the total capacity of the volumetric flask previously filled with the dry sample is reached and the volume

20

required for this purpose is noted.

3.4.2. Chemical analysis

The characterization of the sorption complex, pH, available phosphorus and organic carbon was carried out following the methodologies described in Embrapa (1997 and 1999). The pH values were determined in water and 1M KCl, in a 1:2.5 ratio. The calcium and calcium plus magnesium contents were determined using KCl 1M as the extracting solution and by complexing with EDTA. The magnesium content was obtained by difference. The potassium and sodium contents were obtained after extraction with Mehlich 1 solution (HCl 0.05M+ H2SO4 0.0125M) and determined by flame photometry. Exchangeable aluminum was determined by volumetric titration with sodium hydroxide after extraction with 1M KCl solution. Potential acidity (H+Al) was obtained by extraction with 0.5M calcium acetate pH 7.1 - 7.2 and determined by alkalimetric titration with 0.025M NaOH.

Available phosphorus was also extracted with Mehlich 1 acid solution and determined spectrophotometrically by reducing molybdate with ascorbic acid. Organic matter was estimated by converting the organic carbon content in the soil, determined by the organic matter oxidation method (O.M.) using potassium dichromate in a sulphuric medium, using the ratio O.M. = organic carbon x 1.724.

The following parameters were estimated using the determinations made: sum of bases or SB value (SB = Na+K+Ca+Mg); cation exchange capacity or T value (T= SB + H+Al); base saturation index or V value [V= (SB/T * 100)]; aluminum saturation index or m value [m= (Al/SB+Al)*100] and sodium saturation percentage or PST [PST= (Na/T) * 100].

3.4 Determining soil resistance to penetration

Soil penetration resistance (PR) values were obtained using a Stolf model impact penetrometer (STOLF *et al.*, 1983).

Figure 11 - Stolf model impact penetrometer.

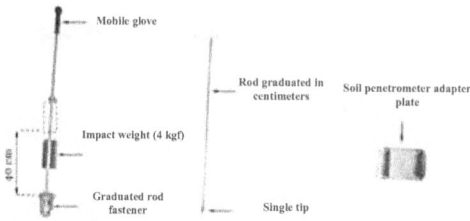

Source: (STOLF *et al.*, 1983)

In this type of penetrometer, a weight of 4 kgf is released from a height of 40 cm and, when it collides with the clamp of the graduated rod, the system, through the conical tip, penetrates the soil to a depth x from the initial state.

Impact penetration is read on the penetrometer's graduated rod and the results are given in impacts/dm (number of impacts needed to perforate a decimeter of soil). Currently, impact/dm data must be transformed into the international system, i.e. into kgf/cm^{-2} and then into Mpa in order to make better use of them.

The Dutch model equation below (Equation 1) (Stolf, 1991) was used to make the necessary transformations.

$$RP = \frac{(M+m).g}{A} + \left(\frac{f.M.g.H}{10.A}\right).N.$$
(1)

Where:

R is the resistance of the soil to penetration in kgf.cm^{-2} ; M is the impact mass (4 kg, commercial model); m is the mass of the penetrometer body (3.2 kg); g is the acceleration due to gravity; f is the fraction of energy remaining to promote penetration [M/(M+m)]; H is the height of the fall of the impact mass (40 cm); N is the number of impacts per decimeter and A is the area of the base of the fine-tip penetration cone (1.28 cm^2). This can be summarized in equation 2:

$$R(kgf.cm^{-2}) = 5,6 + 6,89.N$$
(2)

And to transform kgf.cm^{-2} to mega pascal (MPa) just use equation 3:

22

$$R(MPa) = 0,0980665 . R(kgf.cm^{-2})$$

(3)

The PRs were obtained at depths of 0-10cm and 10-20cm, respectively.

A representative PR profile was obtained for each sampling site, corresponding to the depth averages of the values obtained from four replicates in each treatment. Soil moisture during the collection period was determined using an electronic soil moisture meter, model HFM2010, which takes an electromagnetic measurement called ISAF (Impedance of Soil at High Frequency), which is proportional to moisture. The equipment digitally reports the moisture values at a depth of 20 cm.

1.6. Determining the water infiltration rate in the soil

The rate of water infiltration into the soil was measured using the concentric ring infiltrometer method (CALHEIROS *et al.*, 2009). This equipment consists of two concentric sheet metal rings (Figure 12) with diameters varying between 16 and 40 cm. Water is applied to both cylinders, maintaining a liquid layer of between 1 and 5 cm. In the inner cylinder, the volume applied at fixed time intervals is measured, as well as the consecutive water levels. The outer cylinder has the purpose of vertically maintaining the flow of water from the inner cylinder, where the infiltration rate is measured.

Soil water infiltration measurements were taken at pre-established time intervals in minutes, with 3 replicates chosen at random in each treatment.

Figure 12 - Concentric ring infiltrometer.

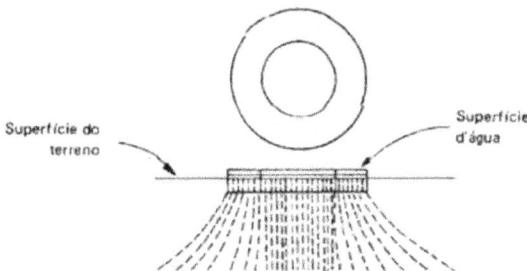

3.7. Determining soil porosity and density

Soil density and microporosity were determined at the Soil Science Department, PICI Campus. Undeformed samples were collected in December 2011, using an Uhland sampler (Figure 13), at depths of 0-10cm and 10-20cm, with three replicates in each treatment at the respective depths, with a total of 18 samples taken for the study.

The samples were wrapped in murim cloth to keep structural changes and water loss to a minimum. In the soil physics laboratory at the Federal University of Ceará - PICI Campus in Fortaleza, the samples were prepared by removing the excess soil on both sides using a spatula and vacuum pump. Subsequently, the samples were placed in a tray, with water gradually added up to the top edge of the cylinder, remaining for a period of 24 hours for saturation and then subjected to 60 cm of water column on the tension table. After this stabilization period, the samples were dried in an oven at 105° C and the weight of the cylinder+pane+alloy was deducted during weighing. Total porosity, microporosity and macroporosity were calculated using the method described in Embrapa (1997). Equation 4 was used to determine microporosity (%).

$$Microporosidade\ (\%) = [\frac{(PA\ 60\ cm - PAS)}{VC}] * 100$$

(4)

Where:

PA = weight of sample at 60 cm; PAS = weight of sample dried at 105°; VC = volume of cylinder.

Equation 5 was used to determine total porosity (%).

$$Porosidade\ total\ (\%) = [\frac{(DP - DS)}{DP}] * 100$$

(5)

Where:

DP = particle density; DS = soil density.

24

Macroporosity was obtained by difference according to equation 6.

Macroporosity (%) = total porosity (%) - microporosity (%)

Figure 13 - Uhland-type sampler with volumetric ring.

Source: Thiago Leite de Alencar

3.8. Statistical analysis

The experimental design was entirely randomized, consisting of three treatments and three replications. The treatments studied were: native forest, pasture and corn in a conventional planting system.

The results obtained were subjected to analysis of variance and the means of the treatments were compared using the Tukey test at 5% significance, using the Statistical Analysis and Experiment Planning System - Sisvar (Version 5.3 Build 77), a computer program developed by the Federal University of Lavras (UFLA).

4. RESULTS AND DISCUSSION

4.1 - Soil classification

4.1.1 - Results of the morphological analysis

The morphological characteristics of the soil profile studied are shown in Tables 1, 2 and 3.

Waxyness was a morphological characteristic that varied between the horizons, since at depth there was an increase in degree and quantity, while in the superficial part it was not found.

The differentiation in color between the horizons within the profile is clear and was classified as abrupt and flat. The soil showed a darker color in the surface horizons, which may be due to the influence of the amount of organic matter, which tends to be found in lower concentration in the deeper levels. Stickiness and plasticity are other morphological attributes that increased with depth, which can be explained by the increase in clay content.

Table 1 - Morphological analysis results.

Horz.	Professor	Relief	Transition between horizons	Structure		Cerosity	Textural class
(cm)							
A1	0 - 17	Plan	Flat/abrupt	Angular and blocks/medium large/moderate	subangular and	Not shown	Franco Arg. Aren.
A2	17 - 31	Plan	Flat/abrupt	Angular and blocks/small medium/weak	sub-angular and	Not shown	Franco Arg. Aren
BA	31 - 65	Plan	Flat/abrupt	Angular and blocks/Small medium/weak	sub-angular and	Moderate/ abundant	Argilo Aren.
Bt1	65 - 100	Plan	Flat/abrupt	Angular and blocks/medium large/moderate	subangular and	Weak	Clay
Bt2	100 - 160	Plan	Flat/abrupt	Angular and blocks/very small moderate	subangular and small	Weak	Loamy
Bt3	160 - 200	Plan	Flat/abrupt	Angular and blocks/small medium/moderate	sub-angular and	Moderate/ abundant	Loamy

Source: Thiago Leite de Alencar

Table 2 - Morphological analysis results.

Horz.	Professor	Soil color		Consistency		Wet	
	(cm)	Dry	Wet	Drought	Wet	Plasticity	Stickiness
A1	0 - 17	7.5YR ¾	7.5YR 2/2	Very hard	Cold	Slightly plastic	Slightly sticky
A2	17 - 31	7.5YR 4/4	7.5YR ¾	Hard	Cold	Slightly plastic	Slightly sticky
BA	31 - 65	7.5YR 5/6	7.5YR 4/6	Very hard	Cold	Slightly plastic	Slightly sticky
Bt1	65 - 100	5YR 5/8	5YR 4/6	Extremely hard	Firm	Slightly plastic	Sticky
Bt2	100 - 160	2.5YR 4/6	5YR 5/8	Hard	Firm	Slightly plastic	Sticky
Bt3	160 - 200	2.5YR 4/6	5YR 5/8	Hard	Firm	Plastic	Slightly sticky

Source: Thiago Leite de Alencar

Table 3 - Morphological analysis results.

Horz.	Professor	Crops / vegetation	Distribution of roots	Pores
(cm)				
A1	0 - 17	Hypoxerophilous Caatinga	Very thin and thick / many	Large and common
A2	17 - 31	Hypoxerophilous Caatinga	Very thin and thick / many	Large and common
BA	31 - 65	Hypoxerophilous Caatinga	Thin and thick/common	Small and medium/common
Bt1	65 - 100	Hypoxerophilous Caatinga	Thin/small	Small and medium/common
Bt2	100 -160	Hypoxerophilous Caatinga	Thick/thin	Small and medium/common
Bt3	160 -200	Hypoxerophilous Caatinga	Thick/thin	Small and medium/common

Source: Thiago Leite de Alencar

1.1.3 - Results of the physical analysis

The results obtained from the physical analysis of the profile are shown in Table 4.

Table 4 - Results of physical analysis.

Horz.	Professor	Particle size composition (g kg^{-1})				Particle density
		Coarse sand	Fine Sand	Silt	Clay	
	(cm)		(g kg^{-1})			
A1	0 - 17	171	456	149	225	2,47
A2	17 - 31	177	386	100	336	2,50
BA	31 - 65	137	344	92	427	2,46
Bt1	65 - 100	128	297	87	488	2,54
Bt2	100 -160	121	310	190	380	2,36
Bt3	160 -200	102	287	317	295	2,30

Source: Thiago Leite de Alencar

1.1.4 - Result of the chemical analysis

The results obtained from the chemical analysis of the profile are shown in Table 5. The profile showed base saturation above 50% in all horizons, showing its eutrophic nature, low to medium sum of bases ranging from 1.71 to 4.80 cmolc. kg^{-1}, medium to low cation exchange capacity ranging from 2.95 to 7.77 cmolc. kg^{-1}, as well as very low,

low and medium levels of calcium (0.2 to 2.1 cmolc. kg^{-1}), good to very good levels of magnesium (1.2 to 2.4 cmolc. kg^{-1}), sodium (0.21 to 0.23 cmolc. kg^{-1}), potassium (0.07 to 0.14 cmolc. kg^{-1}), low levels of aluminum (0.07 to 0.10 cmolc. kg^{-1}) and weak acidity and alkalinity.

Table 5. - Results of chemical analysis.

Hor	pH (1:2.5)		Ca	Mg	In	K	H + Al	Al	S	T	P	PST	V	m	C	MO
	Water	KCl					cmolc kg^{-1}					%			g kg^{-1}	
A1	6,3	5,4	2,1	2,4	0,23	0,07	2,97	0,10	4,8	7,8	6,52	2,9	62	2	12,7	21,9
A2	6,3	5,3	1,0	2,4	0,25	0,07	2,89	0,07	3,7	6,6	4,10	3,8	56	2	11,4	19,6
BA	6,6	5,3	0,8	1,3	0,25	0,10	2,48	0,10	2,4	4,9	2,17	5,1	50	4	5,86	10,1
Bt1	7,0	5,7	0,4	1,6	0,25	0,16	1,90	0,10	2,4	4,3	2,29	5,8	56	4	5,40	9,3
Bt2	7,1	5,9	0,3	1,2	0,22	0,14	1,65	0,10	1,8	3,5	2,53	6,3	53	5	2,99	5,1
Bt3	6,5	5,9	0,2	1,2	0,21	0,10	1,24	0,07	1,7	2,9	3,42	7,1	58	4	2,14	3,7

Source: Thiago Leite de Alencar

1.1.5 Soil classification results

Based on the results obtained from the morphological, physical and chemical analysis of the profile, the soil was classified up to the sixth categorical level, according to the Brazilian Soil Classification System (EMBRAPA, 2006). The soil has low clay activity, a textural ratio of 1.63 between the A and B horizons, together with moderate and abundant cerosity in one or more subhorizons in the upper part of the B horizon. Therefore, it was concluded that the profile under study is classified as typical eutrophic yellow-red Argissolo, medium texture/clay texture, moderate A, mesoeutrophic, very deep, neutral.

4.2 Soil resistance to penetration

The results obtained from the penetration resistance analyses at the 0-10 cm and 10-20 cm depths are shown in Figures 14 and 15 respectively.

Figure 14 - Variation in average PR in the different treatments at a depth of 0 to 10 cm in the soil, Assaré, 2011.

28

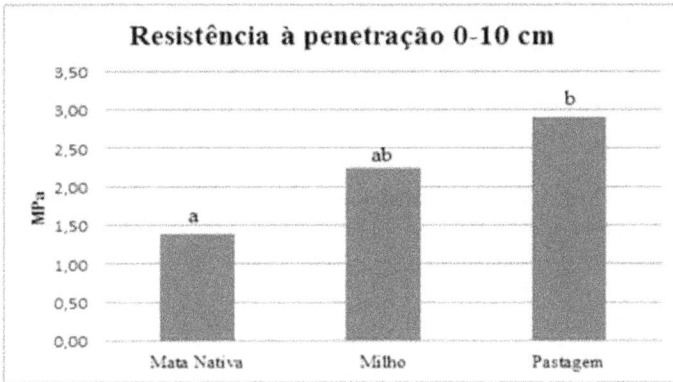

Resistência à penetração 0-10 cm

Source: Thiago Leite de Alencar

Averages followed by the same letters do not differ significantly at 5% by the Tuckey test.

According to figure 14, the treatment with the highest degree of resistance to penetration was the pasture treatment, which differed significantly from the native forest treatment. This can be explained by the large number of animals in the area over a long period of time, which leads to intense trampling and a consequent increase in the degree of surface compaction compared to the other treatments. The values of soil resistance to penetration normally used as critical for full plant growth and development are above 2 MPa (SILVA, *et al.*, 1994; TORMENA *et al.*, 1998; LAPEN *et al.*, 2004; CANARACHE *et al.*, 1990).

The maize treatment had the second highest average, with the presence of animals in the area for a short period of time and the use of machinery in soil preparation and planting being pointed out as possible causes. The native forest, being a preserved environment, had the lowest PR values at a depth of 0-10 cm and was therefore taken as a reference for the study.

Figure 15-Variation of average PR in the different treatments at a depth of 10 to 20 cm in the soil, Assaré, 2011.

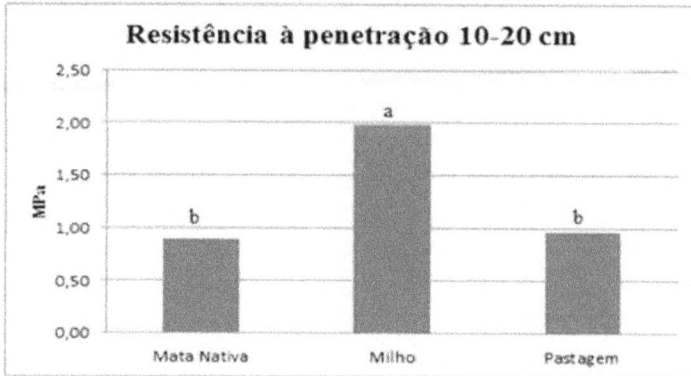

Resistência à penetração 10-20 cm

Averages followed by the same letters do not differ significantly at 5% by the Tuckey test.

According to figure 15, the use of soil with corn had the highest average resistance to penetration in the 10-20 cm layer, differing significantly from the other treatments, which can be explained by the use of soil preparation equipment repeatedly over several years at the same depth, causing an increase in compaction in the subsurface (plow foot). The pasture had a lower average than the corn, due to the fact that the pressure exerted by the animals affects the topsoil more intensely, without showing great effects at depth. The soil with native forest obtained the lowest value among the treatments analyzed in the 10-20 cm layer, agreeing with the results obtained by (ARAUJO, 2010) due to the preservation of the area, keeping the soil's physical characteristics in an optimal state of conservation.

4.3 - Soil moisture

During the penetration resistance sampling period, soil moisture was measured at a depth of 0-20 cm (Table 6). The treatment with native forest had the highest average, due to the influence of vegetation cover and dew on maintaining and retaining soil moisture. The pasture treatment had the second highest moisture content, as it has a certain amount of vegetation which acts to protect the soil, thus preventing greater evaporation. The corn treatment had 13% moisture, due to its use as a conventional crop, leaving the soil less protected.

Table 6 - Moisture content.

30

Treatment	Humidity (%)
Native Forest	21
Corn	13
Pasture	15

Source: Thiago Leite de Alencar

4.4 - Soil density, macroporosity, microporosity

Table 7 - Average values for soil density (Ds), macroporosity, microporosity and total porosity (0-10 cm).

Treatment	Ds (g.cm)$^{-3}$	Macroporosity (%)	Microporosity (%)	Porosity Total (%)
0 - 10 cm				
Native Forest	1,35 b	19,4 a	30,3 a	49,7
Corn	1.45 ab	9,5 b	29,8 b	39,3
Pasture	1,56 a	8,3 b	28,6 b	36,9

Source: Thiago Leite de Alencar

Averages followed by the same letters do not differ significantly at 5% by the Tuckey test.

Table 7 shows an inversely proportional relationship between soil density and macroporosity. At a depth of 0-10 cm, the pasture treatment had the highest average soil density, differing significantly from the woodland treatment. This is due to a reduction in the amount of macropores in the surface layer, generally caused by increased soil compaction due to the high density of animals in the area. The maize treatment did not differ significantly from the pasture due to the presence of animals in the area for a short period of time. Both treatments have average soil density values below 1.65 Mg m^{-3} (g.cm^{-3}), which according to Llanillo *et al.* (2006), in medium-textured soils is restrictive to root growth and, consequently, to the yield of cultivated species. In contrast, the native forest had a lower average soil density and a higher percentage of macropores.

Table 8 - Average values for soil density (Ds), soil macroporosity and soil microporosity (10-20 cm).

Treatment	Ds (g.cm)$^{-3}$	Macroporosity (%)	Microporosity (%)	Porosity Total (%)
10 - 20 cm				
Native Forest	1,30 b	19,6 a	27,6 a	47,2
Corn	1,56 a	8,0 b	28,9 a	36,9
Pasture	1,52 a	9,7 b	30,2 a	39,9

Source: Thiago Leite de Alencar

Averages followed by the same letters do not differ significantly at 5% by the Tuckey test.

Evaluating the 10-20 cm depth (Table 8), the maize treatment had the highest soil density compared to the other treatments, differing significantly from the native forest and, consequently, was the use that most altered the soil's physical characteristics at this depth. Once again, the native forest obtained more adequate values for soil density and macroporosity, in agreement with the results obtained by Matias *et al.* (2009), showing that it was the most sustainable use among the treatments evaluated. Therefore, Ds obtained higher values with intensification of land use, showing similar results (Blainski *et al.*, 2008).

4.5 - Soil water infiltration rate

According to Table 9, we can see that the infiltration of water into the soil was higher in the treatment using native forest, because it has physical characteristics that facilitate the entry of water into the soil, such as the amount of organic matter and plant remains on the soil, reducing the impact of rain drops on the surface, avoiding surface runoff and increasing the infiltration rate, in agreement with the results obtained by (CAVENAGE 1996; SUZUKI *et al.* 2000; SOUZA; ALVES 2003b) . The maize treatment had the second highest water infiltration capacity in the soil because it is an area that is used annually for planting and, consequently, the use of machinery decompresses the soil in the surface layer, favoring the infiltration rate.

Soil use with pasture had the lowest water infiltration capacity in the soil and was significantly different from the other treatments. This may be related to intense animal trampling, compacting the surface layer and reducing its porosity, an important aspect in infiltration. This favors the existence of surface runoff and soil erosion, causing a degradation of its physical quality and, therefore, a significant reduction in its productive capacity.

Table 9 - Average values for soil water infiltration rate (mm/min) and average infiltration rate (mm/h).

	Water infiltration into the soil	
Treatments	Average (Infiltration rate mm/min)	Infiltration (mm/h)
Native Forest	2,3774 b	142,6
Corn	1,7111 b	102,7
Pasture	0,4518 a	27,1

Source: Thiago Leite de Alencar
Averages followed by the same letters do not differ significantly at 5% by the Tuckey test.

32

5. CONCLUSIONS

The soil used for native grazing showed a greater degree of compaction in the surface layer, with higher values of resistance to penetration, soil density and consequently lower macroporosity, with much reduced water infiltration compared to the other treatments.

The subsurface layer of the soil was affected in its physical attributes by the use of corn in conventional planting, where this, compared to the other treatments, obtained values that indicate greater compaction, observed by the reduction in macroporosity and increase in soil density and resistance to penetration.

Evaluating and monitoring the soil's physical characteristics is of fundamental importance, as they act as indicators of quality and sustainability, explicitly showing which uses and managements are responsible for the most intense soil degradation.

The soil was classified as Argissolo Vemelho-amarelo eutròfico tipicos, textura média/textura argilosa, A moderado, mesoeutrófico, muito profundo, neutro.

6. BIBLIOGRAPHICAL REFERENCES

ALBUQUERQUE, J.A.; SANGOI, L.; ENDER, M. Effects of crop-livestock integration on soil physical properties and corn crop characteristics. **Revista Brasileira de Ciências do Solo**, Viçosa, v. 25, p. 717-723, 2001.

ALMEIDA, C.X.; CENTURION, J.F.; FREDDI, O.S.; JORGE, R.F. & BARBOSA, J.C. Pedotransfer functions for the soil resistance to penetration curve. **Revista Brasileira de Ciência do Solo**, Viçosa, v. 32, p. 2235-2243, 2008.

ALVES, M.C. & CABEDA, M.S.V. Water infiltration in a Dark Red Podzolic under two tillage methods, using simulated rainfall with two intensities. **Revista Brasileira de Ciências do Solo**, Viçosa, v.23, p. 753-761, 1999.

AMARO FILHO, J.; ASSIS JÛNIOR, R.N; MOTA, J.C.A. **Fisica do Solo: Conceitos e Aplicaçôes.** Fortaleza: University Press, 2008. 290 p.

ANDREOLA, F.; COSTA, L.M.; OLSZEVSKI, N. Influence of winter mulching and organic and/or mineral fertilization on the physical properties of a Structured Purple Soil. **Revista Brasileira de Ciências do Solo**, Viçosa, v. 24, p.857865, 2000.

ARAUJO, Adriana Oliveira. **Evaluation of soil physical properties and edaphic macrofauna in areas subjected to forest management of native vegetation in the Araripe Plateau.** 2010. Dissertation (Master's in Agricultural Engineering) - Center for Agricultural Sciences, Federal University of Ceará, Fortaleza, 2010.

ARGENTON, J.; ALBUQUERQUE, A.J.; BAYER, C.; WILDNER, P.L. Behavior of attributes related to the shape of the structure of red latosol under tillage systems and cover crops. **Revista Brasileira de Ciências do Solo**, Viçosa, v. 29, p. 425-435, 2005.

ARSHAD, M.A.; LOWERY, B.; GROSSMANN, B. Physical tests for monitoring soil quality. In: DORAN, J.W.; JONES, A.J. (Ed.). Methods for assessing soil quality. Madison: **Soil Science Society of America**, 1996, p.123-141 (SSSA Special publication, 49).

BENGHOUGH, A.G. & MULLINS, C.E. Mechanical impedance to root growth: A review of experimental techniques and root growth responses. **J. Soil Sci**, V.41, p

341358, 1990.

BERGAMIN, C,A.; VITORINO, T. C. A.; FRANCHINI, C. J.; SOUZA, A. M. C.; SOUZA, R. F. Compaction in a dystrophic red latosol and its relationship with maize root growth. **Revista Brasileira de Ciência do Solo**, Viçosa, v.34, p. 681-691, 2010.

BERTOL, I.; GOMES, K.E.; DENARDIN, R.B.N.; ZAGO, L.A.; MARASCHIN, G.E. Soil physical properties related to different levels of forage supply in a natural pasture. **Pesquisa Agropecuâria Brasileira**, Brasilia, v.33, n.5, p.779786, 1998.

BERTOL, I.; ALBUQUERQUE, A.J.; LEITE, D.; AMARAL, J.A.; JUNIOR, Z.A.W. Soil physical properties under conventional tillage and direct sowing in crop rotation and succession, compared to native field. **Revista Brasileira de Ciências do Solo**, Viçosa, v. 28, p.155-163, 2004.

BETTERIDGE, K.; MACKAY, A.D.; SHEPHERD, T.G.; BARKER, D.J.; BUDDING, P.J.; DEVANTIER, B.P. & COSTALL, D.A. Effect of cattle and sheep treading on surface configuration of a sedimentary hill soil. Aust. **J. Soil Res.**, v.37, p.743-760, 1999.

BEUTLER, N.A.; CENTURION, F.J.; SILVA, P.A. Comparison of penetrometers in the evaluation of the compaction of latosols. **Eng. Agric.**, Jaboticabal, v.27, n.1, p.146151, jan./abr. 2007.

BHARATI, L.; LEE; K.H.; ISENHART; T.M. & SCHULTZ, R.C. Soil water infiltration under crops, pasture, and established riparian buffer in Midwest USA. **Agrof. Syst**, v. 56 p. 249-257, 2002.

BLAINSKI, E.; TORMENA, A.C.; FIDALSKI, J.; GUIMARÂES, L.M.R. Quantification of soil physical degradation by means of the soil resistance to penetration curve. **Revista Brasileira de Ciência do Solo**, Viçosa, v. 32, p. 975-983, 2008.

BRAIDA, A.J.; REICHERT, M.J.; VEIGA, M.;REINERT, J.D. Plant residues on the surface and soil organic carbon and their relationship with the maximum density obtained in the proctor test. **Revista Brasileira de Ciências do Solo**, Viçosa, v.30, p.605-614, 2006.

CALHEIROS, M.B.C.; TENÓRIO, C.J.F.; CUNHA, L.X.L.J.; DA SILVA, F.D.; DA SILVA, C.A.J. Definition of infiltration rate for sizing sprinkler irrigation systems. **Revista Brasileira de Engenharia Agricola e Ambiental**, v. 13, n.6, p. 665-670, 2009.

CAMARGO, O.A. & ALLEONI, L.R.F. **Soil compaction and plant development. Piracicaba, Degaspar**, 1997.132p.

CANARACHE, A. Penetrometer - a generalized semi-empirical model estimating soil resistance to penetration. **Soil Tillage Research,** Amsterdam, v.16, p.51-70, 1990.

CARVALHO, C.A.M.; SORATTO, P.R.; ATHAYDE, F.L.M.; ARF, O.; SA, E.M. Maize productivity in succession to green manures in no-till and conventional tillage systems. **Pesquisa Agropecuària Brasileira**, Brasilia, v.39, n.1, p.47-53, 2004.

CARVALHO, J.L.N.;CERRI, C.E.P.; FEIL, B.J.;PICCOLO, M.C.; GODINHO, V.P.; HERPIN, U.; CERRI, C.C. Conversion of cerrado into agricultural land in the southwestern amazon: carbon stocks and soil fertility. **Scientia Agricola**, v. 66, p. 233-241, 2009.

CARVALHO, P.C.F. Access to land, livestock production and ecosystem conservation in the Brazilian Campos biome: The natural grasslands dilemma. Available at: http://www.icarrd.org/en/eventos/tem/STS%20UFGRS%20Biome.pdf. Accessed on: March 22, 2012.

CASA GRANDE, A.A. **Soil compaction and management in sugarcane cultivation.** In: MORAES, M.H.; M.M.L.; FOLONI, J.S.S. (Ed.). Soil physical quality: study methods - soil preparation and management systems. Jaboticabal: FUNEP, p. 150-97, 2001.

CASTRO FILHO, C.; MUZILLI, O. & PODANOSCHI, A.L. Aggregate stability and its relationship with organic carbon content in a dystrophic purple latosol, as a function of planting system, crop rotation and sample preparation methods. **Revista Brasileira de Ciência do Solo**, Viçosa, v. 22 p. 527-538, 1998.

CAVENAGE, A. **Changes in the physical and chemical properties of a Dark Red Latosol under different uses and management.** Ilha Solteira, Universidade Estadual Paulista, 1996. 75p. (Graduation Work)

CECILIO, A.R.; SILVA, D.D.; PRUSKI, F.F.; MARTINEZ, A.M. Modeling soil water infiltration under stratification conditions using the Green-Ampt equation. **Revista Brasileira de Engenharia Agricola e Ambiental**, v. 7, n.3, p. 415-422, 2003.

CICHOTA, R.; LIER, V.J.Q.; ROJAS, L.A.C. Spatial variability of infiltration rate in red clay loam. **Revista Brasileira de Ciências do Solo**, Viçosa, v. 27, p. 789-798, 2003.

COLLARES, L. G.; RINERT, J. D.; KAISER, R. D. Physical quality of the soil in the productivity of the bean crop in an Argissolo. **Revista Brasileira de Ciência do Solo**, Viçosa, v. 41, p. 1663-1674, 2006.

COLLARES, L.G.; REINERT, J.D.; REICHERT, M.J.; KAISER, R.D. Compaction of a latosol induced by machine traffic and its relationship with the growth and productivity of beans and wheat. **Revista Brasileira de Ciência do Solo**, Viçosa, v. 32, p. 933-942, 2008.

CORSINI, C.P. & FERRAUDO, S. A. Effects of cropping systems on soil density, macroporosity and root development of maize in purple latosol. **Pesquisa Agropecuària Brasileira**, Brasilia, v.34, n.2, p.289-298, 1999.

CORREA, J.C. & REICHARDT, K. Effect of time of pasture use on the properties of a Yellow Latosol from Central Amazonia. **Pesquisa Agropecuària Brasileira**, Brasilia, v.30, p.107-114, 1995.

CORRÊA, M.R.; FREIRE, S.G.B.M.; FERREIRA, C.L.R.; FREIRE, J.F.; PESSOA, M.G.L.; MIRANDA, A.M.; MELO, M.V.D. Atributos quimicos de solos sob diferentes usos em perimetro irrigado no semiàrido de Pernambuco. **Revista Brasileira de Ciência do Solo**, Viçosa, v. 33, p. 305-314, 2009.

DO VALE, C. J.; NETO, F. R.; SILVA, L.S.P. selection index for maize cultivars with dual aptitude: mini-maize and green maize. **Bragantia**, Campinas, v. 70, n. 4, p.781-

787, 2011.

EMBRAPA. National Soil Research Center. **Brazilian soil classification system.** 2. ed. Rio de Janeiro: Embrapa Solos, 2006. 306 p.

EMBRAPA. **National Soil Research Center.** Manual of soil analysis methods. 2 ed. rev. and current. Rio de Janeiro: EMBRAPA, 1997. 212p.

FILHO, T. J. & RIBON, A. A. Soil resistance to penetration in response to the number of samples and type of sampling. **Revista Brasileira de Ciência do Solo**, Viçosa, v. 32, p. 487-494, 2008.

FLOWERS, M.D. & LAL, R. Axle load and tillage effects on soil physical properties and soybean grain yield on a Molic Ochraqualf in Northwest. **Soil Tillage Res**, v. 48 p. 21-35, 1998.

FREITAS, P. L., Physical and biological aspects of the soil. In: LANDERS, J.N. Ed. **Experiências de Plantio Direto no Cerrado. Goiânia**: APDC, 1994. p.199-213. 261p.

GAMA-RODRIGUES, E.F. & GAMA-RODRIGUES, A.C. Microbial biomass and nutrient cycling. In: SANTOS, G.A.; SILVA, L.S.; CANELLAS, L.P. & CAMARGO, F.A.O., eds. **Fundamentals of soil organic matter tropical and subtropical ecosystems.** 2.ed. Porto Alegre, Metrópole, 2008. p.159-170.

GIRARDELLO, C.V.; AMADO, CJ.T.; NICOLOSO, S.R.; HORBE, N.A.T.; FERREIRA, O.A.; TABALDI, M.F.; LANZANOVA, E.M. Changes in the physical attributes of a red latosol under no-tillage induced by different types of scarifiers and soybean yield. **Revista Brasileira de Ciências do Solo**, Viçosa, v. 35, p. 2115-2126, 2011.

GOMES, A. & PENA, Y. A. Characterization of compaction using a penetrometer. **Lavoura Arrozeira**, Porto Alegre, v.49, n.1, p.18-20, 1996.

GONÇALVES, E.N. **Ingestive behavior of cattle and sheep on natural pasture in the Central Depression of Rio Grande do Sul.** Porto Alegre, Federal University of Rio Grande do Sul, 2007. 138p. (Doctoral thesis)

GUIMARAES, M. C.; STONE, F. L.; MOREIRA, A. A. Soil compaction in bean cultivation. II: effect on root and shoot development. **Revista Brasileira de Engenharia Agricola e Ambiental**, v. 6, p. 213-218, 2002.

IBGE 2005 **Map of Biomes and Vegetation. Brazilian Institute of Geography and Statistics. Rio de Janeiro.** Available at: http://www.ibge.gov.br Accessed on March 23, 2012.

JORGE, J.A.; CAMARGO, O.A.; VALADARES, J.M.A.S. Physical conditions of a Dark Red Latosol four years after the application of sewage sludge and limestone. **Revista Brasileira de Ciências do Solo**, Viçosa, v. 15, p. 237-240, 1991.

. **Exploratory survey: soil reconnaissance of the State of Cearà**. Recife: SUDENE/EMBRAPA, 1973 (Bol. Téc. 28, Série pedològica, 16).

KARLEN, D.L.; MAUSBACH, M.J.; DORAN, J.W.; CLINE, R.G.; HARRIS, R.F. & SCHUMAN, G.E. Soil quality: A concept, definition and framework for evaluation. **Soil Sci. Soc. Am. J.**, 61:4-10, 1997

KAMIMURA, M.K.; ALVES, C.M.; ARF, ORIVALDO; BINOTTI, S.F.F. Physical properties of a red latosol under upland rice cultivation in different soil and water managements. **Revista Brasileira de Ciência do Solo**, Viçosa, v. 68 p. 723-731, 2009.

LAIA, M. A.; MAIA, S. C. J.; KIM, E. M. Use of an electronic penetrometer to evaluate the resistance of soil cultivated with sugarcane. **Revista Brasileira de Engenharia Agricola e Ambiental**, v. 10, p. 523-530, 2006.

LAPEN, D.R.; TOPP, G.C.; GREGORICH, E.G. & CURNOE, W.E. Least limiting water range indicators of soil quality and corn production, Eastern Ontario, Canada. **Soil Till. Res.**, v. 78, p. 151-170, 2004.

LEAO, P.T.; SILVA, P.A.; MACEDO, M.C.M.; IMHOFF, S.; EUCLIDES, B.P.V. Optimum water interval in the evaluation of continuous and rotational grazing systems. **Revista Brasileira de Ciências do Solo**, Viçosa, v.28, p.415-423, 2004.

LIPIEC, J.; HÂKANSSON, I.; TARKIEWICZ, S.; KOSSOWSKI, J. Soil physical properties and growth of spring barley related to the degree of compactness of two soils. **Soil & Tillage Research**, v.19, p.307-317, 1991.

LLANILLO, R.F.; RICHART, A.; FILHO, J.T.; GUIMARÂES, M.F.; FERREIRA, R.R.M. Evoluçâo de propriedades fisicas do solo em funçâo dos sistemas de manejo em culturas anuais. **Semina: Ciências Agrârias**, Londrina, v.27, n.2, p.205-220, 2006.

MATIAS, R.S.S.; BORBA, A.J.; TICELLI, M.; PANOSSO, R.A.; CAMARA, T.F. Atributos fisicos de um Latossolo Vermelho submetido a diferentes usos. **Rev. Cienc. Agron.**, Fortaleza, v. 40 n. 3, p. 331-338, 2009.

MELO, O.R.; PACHECO, P.E.; MENEZES, C.J.; CANTALICE, B.R.J. Susceptibility to compaction and correlation between physical properties of a neosol under caatinga vegetation. **Caatinga (Mossoró, Brazil)**, v.21, n.5, p.12-17, 2008

MENDES, G.F.; MELLONI, P.G.E.; MELLONI, R. Aplicação de atributos fisicos do solo no estudo da qualidade de áreas impactadas, em Itajubâ/MG. **Cerne**, Lavras, v. 12, n. 3, p. 211-220, 2006.

MONTANARI, R.; CARVALHO, P.M.; ANDREOTTI, M.; DALCHIAVON, C.F.; LOVERA, H.L.; HONORATO, O.A.M. Aspects of bean productivity correlated with soil physical attributes under a high technological level of management. **Revista Brasileira de Ciência do Solo**, Viçosa, v. 34 p. 1811-1822, 2010.

NEVES, C.M.N.N.; SILVA, M.L.N.; CURI, N.; CARDOSO, E.L.; MACEDO, R.L.G.; FERREIRA, M.M. & SOUZA, F.S. Soil quality indicator attributes in an agrosilvopastoral system in the northwest of Minas Gerais State. **Sci. For.**, 74:45-53, 2007.

NIERO, L.A.C; DECHEN, S.C.F.; COELHO, R.M.; DE MARIA, I.C. Visual evaluations as an index of soil quality and their validation by physical and chemical analyses in a dystrophic red latosol with different uses and managements. **Revista Brasileira de Ciência do Solo**, Viçosa, v. 34 p. 1271-1282, 2010.

PACHECO, E.P. & CANTALICE, J.R.B. Path analysis in the study of the effects of physical attributes and organic matter on the compressibility and resistance to penetration of an Argissolo cultivated with sugarcane. **Revista Brasileira de Ciência**

do Solo, Viçosa, v. 35, p. 417-428, 2011.

PANACHUKI, E.; BERTOL, I.; SOBRINHO, A.T.; OLIVEIRA, S.T.P.; RODRIGUES, B.B.D. Soil and water losses and water infiltration in red latosol under management systems. **Revista Brasileira de Ciências do Solo**, Viçosa, v. 37, p. 1777-1785, 2001.

PAULETTO, E. A.; BORGES, J. R.; SOUSA, R. O.; PINTO, L. F. S.; SILVA, J. B.; LEITZKE, V. W. Evaluation of the density and porosity of a gleissolo submitted to different cultivation systems and different crops. **Revista Brasileira de Agrociência**, Pelotas, v.11, n. 2, p. 207-210, 2005.

PIRES, F.L.; ROSA, A.J.; TIMM, C.L. Comparison of soil density measurement methods. **Acta Scientiarum Agronomy**, Maringà, v. 33, p. 161-170, 2011.

POTT, A.C. & DE MARIA, C.I. Comparison of field methods for determining basic infiltration velocity. **Revista Brasileira de Ciências do Solo**, Viçosa, v. 27, p. 19-27, 2003.

REICHERT, J.M.; SUZUKI, L.E.A.S.; REINERT, D.J.; HORN, R. & KANSSON, I.H. Reference bulk density and critical degree-of-compactness for no-till crop production in subtropical highly weathered soils. **Soil Tillage Res.**, v.102, p 242-254, 2009.

REINERT, J.D.; ALBUQUERQUE, A.J.; REICHERT, M.; AITA, C.; ANDRADA, C.M.M. Critical soil density limits for root growth of cover crops in red clay loam. **Revista Brasileira de Ciência do Solo**, Viçosa, v. 32 p. 1805-1816, 2008.

RIBON, A.A. & FILHO, T.J. Estimation of the mechanical resistance to penetration of a red latosol under perennial cropping in the north of the state of Paranâ. **Revista Brasileira de Ciências do Solo**, Viçosa, v. 32, p. 1817-1825, 2008.

ROTH, C.H. & PAVAN, M.A. Effect of lime and gypsum on glay dispersion and infiltration in samples of a Brazilian Oxisols. **Geoderma**, v. 48, p. 351-361, 1991.

SANTOS, R.D.; LEMOS, R.C.; SANTOS, G.H.; KER, J.C.; ANJOS, L.H.C. **Manual de descriçâo e coleta de solo no campo**. 5th ed. Viçosa. SBCS, 2005. 100p

SANTOS, R. J.; BICUDO, J. S.; NAKAGAWA, J.; ALBUQUERQUE, W.A.; CARDOSO, L.C. Atributos quimicos do solo e produtividade do milho afetados por corretivos e manejo do solo. **Revista Brasileira de Engenharia Agricola e Ambiental**, v. 10, n.2, p. 323-330, 2006.

SHUKLA, M.K.; LAL, R. & EBINGER, M. Determining soil quality indicators by factor analysis. **Soil Till. Res.**,v. 87, p.194-204, 2006.

SILVA, P. A.; TORMENA, A. C.; FIDALSKI, J.; IMHOFF, S. Pedotransfer functions for water retention curves and soil resistance to penetration. **Revista Brasileira de Ciência do Solo**, Viçosa, v. 32, p. 1-10, 2008.

SILVA, S. F. **Aspectos quimicos e morfológicos de solos em uma toposequência na Ilha de Santiago - Cabo Verde - Africa.** 2010. 48p. Course completion work, monograph (Graduation in agronomy) - Center for Agricultural Sciences, Federal University of Cearà, Fortaleza - CE.

SILVA, A.P.; KAY, B.D. & PERFECT, E. Characterization of the least limiting water range. **Soil Sci. Soc. Am. J.**, v. 58, p. 1775-1781, 1994.

SOARES, N.L.J.; ESPiNDOLA, R.C.; CASTRO, S.S. Physical and morphological changes in soils cultivated under traditional management systems. **Revista Brasileira de Ciência do Solo**, Viçosa, v. 29, p. 1005-1014, 2005.

SOUZA, O.J.W. & MELO, J.W. Organic matter in a latosol subjected to different corn production systems. **Revista Brasileira de Ciências do Solo**, Viçosa, v.27, p.1113-1122, 2003.

SOUZA, Z.M. & ALVES, M.C. Water movement and resistance to penetration in a Cerrado dystrophic red latosol under different uses and management. **R. Bras. Eng. Agric. Amb.**, 7:18-23, 2003b.

STOLF, R.; FERNANDES, J.; FURLANI NETO, V.L. **Recomendações para uso de penetrômetro de impacto, modelo lAA/Planalsucar- Stolf**. Sao Paulo, MIC/IAA/PNMC - Planasulcar, p. 8, 1983 (Impact Penetrometer Series, BT1).

STOLF, R. Theory and experimental test of formulas for transforming impact penetrometer data into soil resistance. **Revista Brasileira de Ciência do Solo, Campinas**, v.15, n.2, p.229-35, 1991.

STOLF, R.; THURLER, M.A.; BACCHI, S.O.O.; REICHARDT, K. Method to estimate soil macroporosity and microporosity based on sand content and bulk density. **Revista Brasileira de Ciências do Solo**, Viçosa, v. 35, p. 447-459, 2011.

SUZUKI, L.E.A.S.; ALVES, M.C. & HIPÓLITO, J.L. Changes in water infiltration in a Red-Yellow Latosol in the northwest of the State of São Paulo under conventional tillage. R. Iniciaçao Cient., 2:57-63, 2000.

SUZUKI, S.A.E.L.; REICHERT, M.J.; REINERT, J.D.; LIMA, R.L.C. Degree of compaction, physical properties and crop yield in Latossolo and Argissolo. **Pesquisa Agropecuâria Brasileira**, Brasilia, v.42, n.8, p.1159-1167, 2007.

TAVARES FILHO, J.; BARBOSA, C. M. G.; GUIMARÂES, F. M.; FONSECA, B. C. I. Resistance of soil to penetration and development of the root system of maize (*zea mays*) under different management systems in a purple latosol. **Revista Brasileira de Ciência do Solo**, Viçosa, v. 25, p. 725-730, 2001.

THIMÓTEO, S.M.C.; BENINNI, Y.R.E.; MURATA, M.I.; FILHO, T.J. Changes in porosity and density of a dystrophic red latosol in two soil management systems. **Acta Scientiarum**, Maringà, v. 23, n. 5, p. 1299-1303, 2001.

TORMENA, C.A. & ROLOFF, G. Dynamics of penetration resistance of a soil under no-tillage. **Revista Brasileira de Ciência do Solo**, Viçosa, v. 20, n. 2, p. 333339, 1996.

TORMENA, C.A.; SILVA, A.P. & LIBARDI, P.L. Characterization of the optimum hydric interval of a purple latosol under no-tillage. **Revista Brasileira de Ciência do Solo**, Viçosa, v. 22, p.573-581, 1998.

TREIN, C.R.; COGO, N.P.; LEVIEN, R. Soil preparation methods in corn cultivation and clover reseeding in the oat + clover / corn rotation, after intensive grazing. **Revista Brasileira de Ciências do Solo**, Viçosa, v.15, p.105-111, 1991.

TSEGAYE, T. & HILL, R.L. Intensive tillage effects on spatial variability of soil physical properties. **Soil Science**, v.163 p.143-154, 1998.

VIEIRA, L.M. & KLEIN, A.V. Physico-hydric properties of a red Iatosol submitted to different management systems. **Revista Brasileira de Ciências do Solo**, Viçosa, v.31, p. 1271-1280, 2007.

9 786208 269258